X-TREME FACTS: ENGI

VEHICLES

by Raymond Bergin

Minneapolis, Minnesota

Credits:

Cover and title page, Pan_photo/Shutterstock.com; 4 Acdixon/Creative Commons; 4 left, Albert Kretschmer and Dr. Carl Rohrbach/Public Domain; 5 top, ddisq/Shutterstock; 5 top middle, 8 bottom, 16 top left, 18 bottom left, 18 bottom middle, 23 bottom far left, 25 middle right, LightField Studios/Shutterstock; 5 top right, dawarwickphotography/Creative Commons; 5 middle, Edward S. Curtis/Public Domain; 5 bottom, John T. Daniels (restoration & colorization by Wright Stuf)/Creative Commons; 5 bottom right, NASA/Public Domain; 6 top, Ewa Studio/Shutterstock; 6 top middle, Magnus Manske/Creative Commons; Thomas Schoch/Creative Commons; 6 bottom, Thomas Schoch/Creative Commons; 7 top, Charles Marion Russell/Public Domain; 7 middle, Carl Rasmussen/Public Domain; 7 bottom, pedrik/Creative Commons; 7 bottom left, Pedro Lira Rencoret/Public Domain; 7 bottom middle,paffy/Shutterstock; 7 bottom right, R. M. Latzke/GNU Free Documentation License; 8 top, National Maritime Museum, Greenwich, London/Public Domain; 9 top, Anonymous. Colourization work by Thomas Schmid/Public Domain; 9 middle, Nickolay Khoroshkov/Shutterstock; 9 bottom, Bryan Busovicki/Shutterstock; 9 bottom left, MidoSemsem/Shutterstock; 9 bottom right, Izabela Magier/Shutterstock; 9 bottom right top, 19 middle, Prostock-studio/Shutterstock; 10 top, 2Pho7oAndres/Shutterstock;10 top middle, Everett Collection/Shutterstock; 10 bottom, Wide World Photos/Public Domain; 11 top, aappp/Shutterstock; 11 middle, Florian Fèvre from Mobilys/Creative Commons; 11 bottom, Kapi Ng/Shutterstock; 11 bottom left, Karel Bartik/Shutterstock; 12, Anton Gvozdikov/Shutterstock.com; 12 left, Flystock/Shutterstock; 13 top, Pinky-Winky/Shutterstock; 13 top upper left, stanga/Shutterstock; 13 top left, ShotPrime Studio/Shutterstock; 13 top right, Olga Oliva/Shutterstock; 13 middle, Heribert Bechen/Creative Commons; 13 bottom, homydesign/Shutterstock; 14 top, Xocolatl/Creative Commons;14 middle, Public Domain;14 bottom, ModelTMitch/Creative Commons; 15 top, RossHelen/Shutterstock; 15 top left, Public Domain; 15 top right, MikeDotta/Shutterstock.com; 15 bottom, chuyuss/Shutterstock; 16 top, Ubahnverleih/Creative Commons; 16 top right, William Moss/Shutterstock; 16 bottom, ABCDstock/Shutterstock; 16 bottom left, Culture Coventry Trust 16 bottom left,/Creative Commons; 16 bottom middle, Viorel Sima/Shutterstock; 16 bottom right, Vauxford/Creative Commons; 16 bottom far right, LifetimeStock/Shutterstock; 17 top, Shedlum/Creative Commons; 17 top left, Roman Samborskyi/Shutterstock; 17 top middle, photomaster/Shutterstock; 17 bottom, Pan_photo/Shutterstock.com; 18 top, haroldguevara/Shutterstock; 18 bottom, tcharts/Shutterstock; 18 bottom right, Ksenija Toyechkina/Shutterstock; 19 top, Konstantin Zaykov/Shutterstock; 19 top left, AerialVision_it/Shutterstock; 19 bottom, WR studio/Shutterstock; 20 top, Diagram Lajard/Creative Commons; 20, ansharphoto/Shutterstock; 20 left, Wellcome Images/Creative Commons; 20 middle, Volodymyr Burdiak/Shutterstock; 20 right, Marzolino/Shutterstock; 21 top, Wright Brothers/Public Domain; 21 middle, M101Studio/Shutterstock.com; 21 middle left, paffy/Shutterstock; 21 middle right, fizkes/Shutterstock; 21 bottom, Viacheslav Lopatin/Shutterstock; 21 bottom center, Richard Vandervord/Creative Commons; 22 top, NASA/Public Domain; 22 middle, NASA Neil A. Armstrong/Public Domain; 22 bottom, NASA/David Scott/Public Domain; 23 top, NASA/JPL-Caltech/Public Domain; 23 middle, 3Dsculptor/Shutterstock; 23 bottom, SpaceX/Creative Commons; 23 bottom middle, Roman Samborskyi/Shutterstock; 23 bottom right, David Steele/Shutterstock; 24 top, Juergen_Wallstabe/Shutterstock.com; 24 top left, Antonio Guillem/Shutterstock; 24 top right, Jeka/Shutterstock; 24 bottom, Hideki Kimura, Kouhei Sagawa/Creative Commons; 25 top, Tobias Klaus (toktok)/Creative Commons; 25 top left, Kuznetsov Dmitriy/Shutterstock; 25 middle, The pods of the personal rapid transit system by David Smith/Creative Commons; 25 bottom, Igor Karasi/Shutterstock.com; 26 top, karamysh/Shutterstock; 26 middle, BearFotos/Shutterstock; 26 bottom, Yauhen_D/Shutterstock; 27 top, Catarina Belova/Shutterstock; 27 top right, IFCAR/Public Domain; 27 middle, andrey_l/Shutterstock; 27 bottom, HelloRF Zcool/Shutterstock; 27 bottom right, Alex Butterfield/Creative Commons; 28 top left, Giorgio Galeotti/Creative Commons; 28 bottom far left, DuraStor consortium/Creative Commons; 28 bottom left, Ristomets/Creative Commons; 28 top right, F16-ISO100/Shutterstock; 28–29, Austen Photography

Bearport Publishing Company Product Development Team

President: Jen Jenson; Director of Product Development: Spencer Brinker; Senior Editor: Allison Juda; Editor: Charly Haley; Associate Editor: Naomi Reich; Senior Designer: Colin O'Dea; Associate Designer: Elena Klinkner; Product Development Assistant: Anita Stasson

Produced for Bearport Publishing by BlueAppleWorks Inc.
Managing Editor for BlueAppleWorks: Melissa McClellan
Art Director: T.J. Choleva
Photo Research: Jane Reid

Library of Congress Cataloging-in-Publication Data is available at www.loc.gov or upon request from the publisher.

ISBN: 979-8-88509-169-5 (hardcover)
ISBN: 979-8-88509-176-3 (paperback)
ISBN: 979-8-88509-183-1 (ebook)

For more information, write to Bearport Publishing, 5357 Penn Avenue South, Minneapolis, MN 55419.
Printed in the United States of America.

Contents

Going Places 4
Let's Sail Away! 6
Shipping Out 8
All Aboard! 10
Pedal Power! 12
Hit the Road 14
Wild Wheels 16
Riding Solo 18
Taking Flight 20
Super Spacecraft 22
Going Green 24
Tomorrow's Transportation 26

Balloon-Powered Car 28
Glossary 30
Read More 31
Learn More Online 31
Index 32
About the Author 32

Going Places

For hundreds of thousands of years, the only way people could get from one place to another was by walking. But humans are always trying to find better ways to do things. Once people began making simple boats and wheels, our tired feet got a break as feats of **engineering** took over! Canoes and carts led to rowboats and bicycles. Then, we figured out how to make planes, trains, and automobiles. Today, we're even taking trips into space!

Wheels were **invented** more than 5,000 years ago in Mesopotamia. **They were made of wood.**

At first, wheels were used to shape clay. But then, someone decided to put them on carts to make moving things easier.

The first car was powered by steam and traveled at about 2 miles per hour (3 kph).
BUT WE'RE ALREADY GOING FULL STEAM AHEAD!
I THINK THEY WANT US TO GO FASTER.
HONK!!!
HONK!!!
175
The earliest boats were canoes carved out of tree trunks.
The first airplane flight lasted only 12 seconds. The plane traveled 120 feet (37 m).
I KNOW! MAYBE YOU SHOULD TAKE A CAR INSTEAD.
UGH, WE'LL NEVER GET THERE AT THIS RATE!
The fastest that humans have ever traveled was in the Apollo 10 spacecraft. It went 24,790 miles per hour (39,897 kph) on its flight back to Earth from the moon.

Let's Sail Away!

Back when walking was the only form of **transportation**, what did people do when they reached rivers or lakes? They built boats, of course! At first, simple canoes and rafts carried people over water. Then, about 5,000 years ago, the ancient Egyptians built bigger boats. These boats used sails to catch wind, which pushed them across the water. Soon, people made even larger vessels to sail across oceans. Let's hop in a boat and see where we can go!

Egyptian sailboats also had huge oars so rowers could keep the boats moving when the wind died.

The largest Egyptian boats could carry as many as 4,000 people.

Chinese sailboats called junks were the first to include **rudders**, which help people steer boats.

Native American and First Nations people in the Pacific Northwest region built long canoes used for traveling, fishing, and even racing.
ROW FASTER!
DON'T BOTHER! THE RACE IS OVER. WE'VE WON!
Viking longships were very thin and light, allowing them to sail long distances quickly.
In 1522, a Spanish sailing ship completed the first around-the-world trip.
LET'S GO FOR A WORLD CRUISE.
IN THAT OLD THING? I DON'T THINK SO!
Sailing ships haven't disappeared—they've just gotten bigger! The modern Royal Clipper is 439 ft (135 m) long and has 42 sails.

Shipping Out

For thousands of years, ships could move across water only with the help of wind or rowers. But in the 1800s, steam power hit the scene and began to replace sails. How did this new kind of power work? Water was boiled to create steam. Then, the steam was directed to a motor, which turned a paddle wheel or propeller to push the ship forward. Ships had a new way to move!

The first steamship was built in 1783. On its first trip, the **boiler** leaked and the ship broke apart!

In 1819, a steamship crossed the Atlantic Ocean for the first time. It took a month to sail from the United States to England.

While steam was replacing sails, **iron** and **steel** ships began replacing wooden ones.

The *Titanic* was the largest steamship in the world in 1912. That is, before it hit an iceberg and sank.
IT'S TIME TO GET ON THE SHIP!
UM . . . I THINK WE SHOULD WAIT FOR THE NEXT ONE . . .
Many steamships were destroyed during World War II (1939–1945). After the war, ships powered by **diesel** fuel took over.
Today, some **freight** ships are as long as four football fields.
Modern cruise ships have made traveling across water more fun. They have movie theaters, water parks, ice rinks, and basketball courts.
LET'S HIT THE WATER PARK!
NAH, I'M GOING TO CHECK OUT THE ICE RINK!

All Aboard!

While steamships were changing the way people traveled across water, steam power was at work on land, too. In the early 1800s, inventors put steam engines on wheels. The steam moved **pistons** that got the wheels going along a set of railroad tracks. This vehicle was called a locomotive, and it pulled railcars filled with passengers and goods. Train travel was born!

Early wooden and iron railroad tracks broke easily under heavy loads. In the 1860s, stronger steel rails began to replace them.

Beginning in the 1950s, most trains stopped steaming! They were powered by diesel fuel or electricity.

In 1830, the Baltimore and Ohio Railroad became **the first railroad to carry both freight and passengers.**

Trains have gotten faster and faster. **Some trains today can move at more than 350 miles per hour (563 kph)!**

The largest freight train ever had 375 cars and was 2.5 miles (4 km) long!

In 1994, trains first sped through the world's longest underwater train tunnel, which connects England and France.

The tip of Japan's superfast bullet train is modeled after the pointy beak of a kingfisher bird.

Pedal Power!

At the same time that steam trains began riding the rails, people started hitting the roads with bicycles. The first bike was made in Germany in 1817. It was wooden with iron wheels and no pedals! Riders kept their bikes moving by pushing their feet off the ground. It was 36 years before another German inventor added pedal power. What else has happened in the world of bikes?

Early bicycles were nicknamed hobby-horses. *Giddyup!*

The first bicycle weighed almost 50 pounds (23 kg). Most bikes today weigh half that.

The iron wheels of early bicycles made for a bumpy ride. This earned them the nickname bone-shakers.

In 1868, the first motorized bicycles—or motorcycles—revved their steam-powered engines.

In the late 1970s, sturdier frames and wider tires **allowed the first mountain bikes to take off-road adventures.**

BMX bikes—designed for off-road racing and stunt tricks—first appeared in the early 1970s in California.

Hit the Road

It wasn't long before riders of hobby-horses and bone-shakers had to share the road with a new kind of vehicle—the automobile. Steam-powered cars appeared as early as 1769. But in 1885, German inventor Carl Benz designed a gas-powered engine and placed it on a car frame. Soon, motorized cars were everywhere!

The first gas-powered automobile hit a top speed of about 10 miles per hour (16 kph).

Inventor Henry Ford found a way to make lots of cars quickly. By 1927, a new car came off his factory's **assembly** line every 24 seconds!

Most early cars came in one color—black.

In 1896, the first speeding ticket was given to a driver going 8 miles per hour (13 kph) in a 2 miles per hour (3 kph) zone.
HOW FAST WAS I GOING, OFFICER?
FAST ENOUGH THAT I HAD TO WALK BRISKLY TO CATCH UP TO YOU!

Modern cars are made of about 30,000 parts.
There are about 1.2 billion cars in the world today. By 2040, there may be 2 billion!
UGH, THIS IS SO BUSY!
YEAH, DOES THERE NEED TO BE SO MANY OF US?
I'M ONE IN A BILLION!

Wild Wheels

Today's cars share crowded highways with huge buses, freight trucks, and many other vehicles. There are some really wild sets of wheels on the road! Tiny cars zip through traffic, while long limousines give riders lots of luxury. Some vehicles are even ready to go off the roads. Let's buckle up and test drive some wacky wheels!

DUKWs were designed to move military troops over land and water. Now, they're used for water rescues and sightseeing tours.

Riding Solo

Those vehicles aren't the only wild rides you could take. Would you want to try riding a floating skateboard or bike? Or zooming across the ground with one wheel and a motor? There are so many unusual personal transportation vehicles. What other ones could you take on a solo trip?

Hover bikes can move 10 ft (3 m) above the ground and go 45 miles per hour (72 kph).

Hop on an electric unicycle that **can go 45 miles per hour (72 kph).**

Hoverboards look like skateboards without wheels. They use strong magnets or electric motors and fans to float above the ground.

Snowed in? A snowmobile's rear tank-like treads and front sled-like **runners** power through the ice and snow.

Scooters with electric motors let people go as fast as **75 miles per hour (121 kph).**

Personal watercraft look like ocean-going snowmobiles. Some can even go on 100-mile (160-km) cruises!

Taking Flight

Ships, automobiles, and all those other wild vehicles help people travel over water and land. But what about taking to the skies? People began experimenting with flight as early as the 800s CE, when inventors built hang gliders. But it wasn't until 1903 that brothers Wilbur and Orville Wright were able to create a powered machine that could stay in the air and be steered. They made the first airplane! Time to soar through the sky . . .

In the early 1500s, artist Leonardo da Vinci **sketched plans for a flying machine after studying the wings of birds.**

When a hot-air balloon with two passengers left the ground in 1783, **it became the first flying machine to carry people.**

The Wright Brothers' early planes could hold only one person. **That person had to lie on their stomach out on the wing!**

Today's largest passenger planes can hold 500 people on two levels. They have restaurants, showers, and beds.

COOL! I'M GOING TO GO GET SOMETHING TO EAT!

I'M GOING TO HIT THE SHOWERS BEFORE BED.

The Concorde airplane flew faster than the speed of sound. It took passengers from France to New York in just 3 1/2 hours.

I NEED TO GO TO THE BATHROOM!

CAN'T YOU WAIT? WE'LL BE IN NEW YORK SOON!

Paris, France

New York City

Atlantic Ocean

Super Spacecraft

Leaving the ground and flying through the air is a pretty amazing feat. But what about leaving the planet and flying through space? For thousands of years, people had looked up at the night sky and imagined traveling among the moon and stars. In 1961, that finally happened. Russian astronaut Yuri Gagarin climbed into a small, round spacecraft attached to powerful rockets and blasted into space! What other vehicles have helped us explore space?